THE SE LAW OF ATTRACTION

Ultimate Guide for Absolute

Beginners On How to

Accomplish All Your Goals and

Dreams

© Copyright 2015

Disclaimer

The information provided in this book is designed to provide helpful information on the subjects discussed. The author's books are only meant to provide the reader with the basics knowledge of the topic in question, without any warranties regarding whether the reader will, or will not, be able to incorporate and apply all the information provided. Although the writer will make his best effort share her insights, the topic in question is a complex one, and each person needs a different timeframe to fully incorporate new information. Neither this book, nor any of the author's books constitute a promise that the reader will learn anything within a certain timeframe.

Gregory Garcia

 Gregory Garcia is a self-defined lifetime learner. He devoted his life to helping others to achieve their dreams and becoming the best version of themselves.

From a young age, Garcia was always fascinated with optimal living and how to live a balanced life. Due to his genuine appreciation to helping others. Garcia began to focus on learning everything he can to achieve his dreams. Along the process he became hugely passionate to helping others by sharing his knowledge through seminars and videos.

Message from Gregory Garcia:

As my gift from me to you for purchasing this book. Download this Free Manual to Learn **How to Literally Force the Universe Give You Everything**. Combine with the content in this book and the manual and you will be astound by the results you will get from it.

To download the "Manifestation Miracle" go this link:

http://bit.ly/1OVXgo6

Contents

Introduction

I want to thank you and to congratulate you for downloading the book, *"The Secret Law of Attraction"*.

This book will act as an authentic guide in ensuring that you understand what the Law of Attraction is and how you can apply it in your life in order to make a difference.

Have you ever wanted something so bad but you kept clouding your mind with so much negativity, then the most probable outcome would have been a fail. That is just how life is, be a person who exercises the secret Law of Attraction by building positivity in all their thoughts and feelings. There are times in life when everything just goes wrong and we tend to blame it on other aspects e.g. fate. What we forget is that our life is a mirror of the thoughts that we think. We all operate through the Law of Attraction and as humans we have an advantage of the mind which can discern, we are therefore able to choose and control our thoughts. We are the ones with the power within that gives us the ability to mould our lives with our minds.

To realize its effectiveness you have to make the Law of Attraction part and parcel of your life. Whatever desires you strongly feel that you have to achieve are all a possibility, all you need is that inner drive that gives you an inner assurance and motivation of success. You being able to apply the secret

Law of Attraction in your life will help you enhance it completely. Be open to developing new streams of thoughts in order for you to attract your goals and dreams.

This book aims at opening you up to a new realization on how to make a difference in your life. Its deepest emphasis lies on the ability to use our thoughts to achieve our dreams. It has been made an interesting read for you with detailed information that will be of so much advantage to you and you are assured of having fun as you read through.

Thanks again for downloading this book, I hope you enjoy it!

CHAPTER ONE

UNDERSTANDING THE PRINCIPLES OF THE LAW OF ATTRACTION

The Law of Attraction is a universal law explaining how our lives are controlled by our thoughts. It operates on the basis that we humans have a certain connection with a "universal energy force" and therefore through our thoughts and feelings we are able to manipulate this energy to suit our desires. The Law of Attraction was popularized by Abraham Hicks and it can also be referred to as the law of divine right, the law of good, the law of creation etc.

If we always focus on the negative things, then our life will be full of negativity but if we decide to focus on positive things then we tend to attract the positive into our lives. We can use this power to be what we want to be and get what we want to get. One must believe, act and feel as though they have already achieved whatever they wish for. That strong attraction and positivity is what will land you your dreams. Being able to create awareness of the existence of the Law of Attraction in your life will allow you to control what you attract into your life.

If you want to change your life then you first have to change your thinking. This is where the idea of the power of attraction comes in. This law states that every individual has the ability to attract things into their lives with their thoughts; it's all about the power of positive thinking. If you think positive things then your life will be filled with positivity. It is believed that the more you say something or see something you will believe in it and it in turn imposes itself into your subconscious and it begins to create the reality.

The Law of Attraction is one of the greatest universal laws. It stipulates that whatever is perceived and stored in the mind the universe will provide exactly that directly back to your life. According to this law, we as humans are the ones who draw our life pictures, much like an artist when you don't like the outcome of the picture what do you do? You change it, making yourself have control of the finished art.

It doesn't matter what you are looking to accomplish or achieve, once you have your mind set on it the universe will surely and entirely hand it over to you. All humans are responsible for who they have become, this is because what you are, good or bad, is as a consequence of what you think. Once you realize the secret of the Law of Attraction you will be in a position to shape up your life as you will have the knowledge that will guide you in redefining your own destiny. You will consequently have hope and renewed sense of purpose to make your life what you want it to be.

The Law of Attraction is simply a way of life. It is what we think, what we perceive, what we hope for and what our eyes store as they see the universe that makes our life. Once one understands the Law of Attraction and how it works they will easily change their life in all dimensions. This is because the

Law of Attraction is applicable in all life aspects e.g. health, love, money, success.

This law puts us in a place which demands that in order for us to achieve our desires we need to create and maintain that mindset. Henry Ford explained it out loud for us, *"Whether you think you can, or you think you can't, you are right,"* it is always all defined in your thoughts. Everything that we manifest or get first existed in the mind as a thought and as we kept focusing on it the universe transformed it into reality.

CHAPTER TWO

TIPS ON HOW TO MAKE THE LAW OF ATTRACTION EFFECTIVE

In order to make the Law of Attraction effective in your life there are certain tips that can be of help. This will help you maintain a positive mindset thus allowing room for proper dream manifestation. It is therefore of essence to ensure that as you exercise the Law of Attraction the following should guide you through:

1. The most important thing in enhancing the Law of Attraction in your life is to have strong faith. Faith is necessary in ensuring that you maintain a positive attitude towards life and what you hope to achieve.

2. Always focus on what you want rather than what you don't want. The Law of Attraction works on the basis of our thoughts and it always attracts whatever your mind holds.

3. Do not focus on the how and when of things happening. Just let the universe work its way by letting everything flow.

4. Work on creating the awareness in you that you have the ability to attract what you want.

5. Do and involve yourself with things that inspire you as a way of ensuring that you maintain a positive thought.

6. Always learn to appreciate life and what it has for you. Along with joy and love, the feeling of appreciation highly influences you to always be positive about life.

7. Clarity is also another essential tip in making the Law of Attraction work for you. Be able to define exactly what you want. Clarity of mind contributes to the clarity of passion.

8. Having your goals written down is also a key element in making the Law of Attraction effective in your life it is believed that written goals are easier to achieve.

9. Another way of aligning your mind with the flow of prosperity and joy is adopting an attitude of gratitude.

COMMON MISTAKES PEOPLE MAKE WHEN APPLYING THE LAW OF ATTRACTION

Learning about the Law of Attraction and how it works is one thing, being able to perfectly apply it in your life is another. The basics of the Law of Attraction is not enough for you to grasp, you have to go a step ahead and get to know some of the major mistakes that are normally made in applying it. These are some of the things that normally make people think that the universe is working against them.

1. Letting your mind go ahead of situations by determining ahead of time how and when you will attract what you want. The Law of Attraction works when you allow it to, by just moving with the flow.

2. Allowing yourself and your emotions to be led by external evidence. This puts you in a position of giving up, because

your experience of reality is different from what you are trying to attract.

3. Another common mistake that many people make is starting out too big. When one is finally aware of the power of Law of Attraction they may want to manifest their biggest dreams. Our biggest dreams are always the ones that seem unbelievable and can therefore be challenging to keep up the faith.

4. Insisting and focusing too much on your dream. For one to have their dream come true you must let go of it and allow the universe to bring it back to you. If you keep insisting on it, it will just stress you up instead.

5. Improper use of affirmations. As much as using affirmations is known to be a great way to manifest your dreams the wrong use of it will only make things wrong. Ensure that your affirmation doesn't make you feel guilty, anxious, angry or depressed.

6. Another mistake that people make is not being clear about what they want.

7. There are also those who judge the situation too quickly. This will do nothing but de-motivate you. Things won't work on your timetable as the Law of Attraction is not instant. You have to flow with the flow.

8. Another mistake is getting obsessed with manifesting what you want. The only reason one is attached to what they intend to manifest is due to doubt and lack of trust in your ability to apply the Law of Attraction.

9. Having very little faith is also another mistake that many people make when it comes to the Law of Attraction. We all know that this law works on faith and if one doesn't believe

that something will happen then it never will. Even having the smallest bit of doubt in you will make it very hard for you to manifest your dream.

10. Impatience is also a mistake that may make the Law of Attraction not to be very effective to you. The Law of Attraction is not an ATM machine where you insert your card and immediately get what you want. It works on its own timing and you should therefore exercise the patience required.

CHAPTER THREE

APPLYING THE LAW OF ATTRACTION ON DIFFERENT ASPECTS OF LIFE

As it has already been explained, the Law of Attraction is all about your thoughts and how you can use them to acquire what you want in life. What you think about you create. The Law of Attraction is applicable in all aspects of life some of the areas that are mostly common are as follows:-

1. Money.

2. Health.

3. Love and relationship.

4. Success.

LAW OF ATTRACTION FOR MONEY

The first thing you should put in mind is that there is no limit to the wealth available to everyone on earth, there is no discrimination, in that some are entitled to unlimited wealth and others are not. No one is predestined to a life of hardship as you alone are responsible for the much wealth you have in your life. It is important to let your mind be consumed with thoughts of abundance and wealth.

Your subconscious mind creates your world as you could use it to achieve your goals. Our whole life is a reflection of our mindset, this is because it is determined by what we have programmed in the mind. We are what we think and the internal thoughts we have are what drives us. Napoleon Hill in his book *"Think and Grow Rich"* explained that whatever the mind can conceive and believe it can achieve. If money is what you want then your thoughts should show exactly the same. Don't be a person of self-doubt by keeping telling yourself that you are poor, you will never get money etc. rather use your mind as your strength.

As you embark on your money manifestation journey it helps to know that money is not much different from trying to attract anything else in your life that is among your desires. When you decide that it is money you want then assure yourself that it is money you will get. Discard any thought of doubt and worry. Don't let the not having enough be a factor in deterring you in your quest. Money is a very huge aspect of our lives as we all need it to survive, it is the reason why some people go to the extent of killing, stealing etc. But why do all that evil when you have a very simple and clear path towards gaining the much you want.

If you keep complaining how poor you are you are just pushing away your financial abundance further and further. You should instead embrace the idea of wealth, start living the dream in your mind by believing in its possibility. The best and most important thing when it comes to money that people don't really know is that being stress free and less worried about money is actually what will lead you to that path of success and open you up to a world of possibilities.

You have to strongly believe that with time, no matter your current situation you will always be great financially. The first step in solving a problem is letting go, once you let go of a problem the solution is able to show up. No one was put on this earth to struggle or suffer and we therefore have to understand that we are in control of our thoughts and thus responsible in shaping our destiny. Make not having money a no option in your life and once you begin to change your thinking and feelings towards money then your life will begin to change for the better.

Bill Gates said, *"I always knew I was going to be rich, I don't think I ever doubted it for a minute"* and he is, no doubt about that. That is the kind of mindset and thoughts that we all need to have. Many rich people always had a positive thought that they would be rich and they never for once doubted the fact that they would be. They always expect to have money and thus money constantly flows to them from the universe.

This is therefore the exact mindset that we should all adopt in order to manifest money or any other of our desires. You must begin by living your dream through your actions. This can be for example checking out houses we love, test driving the car of your dreams etc. These are some of the ways that you can easily refocus your mind to all the positivity that life holds.

Things to keep in mind when using the Law of Attraction to attract money

Appreciating where you are

More things will come to you when you start appreciating what you have at the moment. This is of importance as gratitude is one of the requirements in achieving what you want. Appreciating what you already have is just like telling the universe thank you for this and you ask for more. With this signal the universe will bring more to you.

Use affirmations

Affirmations are sentences that are aimed at affecting the conscious and the subconscious mind. The words composed in the affirmations are meant to inspire, energize and motivate. Repeating affirmations will help in strengthening your power of positive thinking which is a requirement in the Law of Attraction. Affirmations are important because of the following reasons:-

1. They are good for motivation.

2. They keep the mind focused on your goals and dreams.

3. They change your way of thinking and behavior.

4. They also make you feel positive, energetic and active thus in a position to transform your thoughts.

Some of the words/sentences that are used as money affirmations include the following:-

1. I am a money magnet.

2. I am now earning a lot of money.

3. Plenty of money is flowing into my life.

Wipe all negative thoughts

As you already know, the Law of Attraction is all about the universe transforming your thoughts into reality. If you fill your mind with negative thoughts then you are prone to get the same negativities in life. If you are working on allowing the Law of Attraction to bring you money then you have to believe in the idea. Don't be doubtful as it is all possible.

Exercise confidence

It is very important for you to be confident in yourself and know that you can have anything that you think of. No matter your current situation or your background, your thoughts can deliver to you what you always dream of. Confidence is all about believing in yourself and if you do the universe will believe in you too.

Having clear goals

Clarity is very essential for you to attain what you want through the Law of Attraction. In the case of money make it clear to your mind that it is money you want and have the faith that you will get it. Having clear goals will help the universe to know exactly what you want and will be therefore delivered.

LAW OF ATTRACTION FOR HEALTH

No single human being enjoys being sick, we all hope for a life devoid of sickness, diseases and weakness as they all lead to an unhappy and uncomfortable life. According to the Law of Attraction, the healthy and aging free life that we all desire lies in our own hands. We all can attract perfect health with the Law of Attraction because with this law you get whatever you focus your attention on. When you consistently fill your mind with negative thoughts your body is affected and prone to diseases, when you have happy and positive thoughts you can't have any diseases thriving in your body and will always live a healthy life.

The body usually responds to thoughts that are impressed upon it from the subconscious mind, stress, anxiety and negative thoughts break down the body. If you live in fear of attracting diseases then you will manifest them. What you should all realize is that by accepting and believing that diseases are inevitable you will be more than attracting them into your life. When you live in fear of getting sick then you are actually making a choice that you accept the possibility of getting sick and your conscious and subconscious mind will read this and direct it to the universe.

If you believe in the idea of diseases being passed genetically and you encrypt that that idea in your mind thinking how some of your family members have been affected by a certain disease and you believe that you will also be a victim then you will. A good way to handle such is by telling your subconscious mind that as much as chances are high of you getting genetically infected you believe you are an exception and your body is strong enough with a very strong immunity and that it will fight against all kinds of diseases. If this is the kind of signal you send to the universe the then the universe will reverse the same back. Many people usually devote their thoughts to what is affecting them physically, emotionally and mentally and these are things they don't want in their life. The universe gives you exactly what you focus on.

According to the Law of Attraction everything in the universe is a component of certain energy and energy attracts same energy and therefore if you focus on the negative energy you will most likely attract the negative into your life. Our bodies will always react to your quality of thoughts and if you have depressing thoughts it affects your immune system impairing it.

Just like any other life aspects, the power of maintaining a healthy and age-resistant body lies in our hands. Our general health is dependent on our thoughts and benefits. The placebo effect demonstrates how important the mind is in the recovery of the body physically. It explains that the mind has the ability to control itself and the body into a situation of total health without any external support. Combining the healing factor of modern medicine with the mind power will speed up a prison's healing process.

The condition of our physical body is always determined by our mindset. We control this knowingly and unknowingly as we go about our daily lives. Stress is known to be the most common cause of any emotional or physical ill-health. When we notice that our bodies aren't okay, we should first work on changing our mindset. It is very essential to focus your thoughts on positivity, love, happiness and abundance in order to keep your body from manifesting sickness.

For the Law of Attraction to work effectively in regards to health, you must be able to focus on several factors as you go about your daily life.

Health affirmations

It is of essence for one to use Law of Attraction health affirmations as it is a good way to offer positive thoughts that allow Law of Attraction manifestation. Affirmations are simple statements in which one claims the desire that you hold. These are very important as they build your confidence and are a way of maintaining a positive attitude. Examples of health affirmations are as below:

1. I am healthy, whole and complete.

2. I am living a long and healthy life.

3. I am filled with energy to do all the daily activities in my life.

Seeing through different eyes

Seeing through different eyes gives you a new perspective from which to view your world. It is a way through which you will be able to see opportunities and appreciate your life. Being able to view the world in the eyes of a different person with different situations will help you appreciate your life and your situation. You couldn't be the one suffering most in the world. There are others who would give anything to have your life just the way it is.

Appreciation

Appreciation is all about looking at the good in you instead of focusing on the problems. By taking your time to appreciate the Law of Attraction will respond with more positive circumstances. Appreciation will help you focus on the good and therefore reducing stress enabling you to live a happy and healthy life.

Believe

It is important for one to always have faith in achievement of perfect health. It is necessary for you to get yourself in a state of belief in order to embrace healing. This far you already know that faith is a requirement in ensuring that the Law of Attraction works for you. If it is good health you want then you should believe that you will get.

Visualization

It is important for you to always visualize having good health. This works in such a way that your mind will be filled with thoughts of good health and sends signals to the universe of the things that you are attracting; the universe will in turn fill you with perfect health.

LAW OF ATTRACTION FOR SUCCESS

The Law of Attraction is at work all the time for everybody. One doesn't really have to know how it works, what is important is for you to know how to use it. There is no ambition or dream too big that we can't think it into reality. We can use the Law of Attraction to our advantage by using the power of thought to transform ourselves into the successful person that we always dream of. It doesn't matter who you are or how bad your current situation is in life, when you want success in life you have to believe in yourself. There are no restrictions to how much one can achieve.

There has been emphasis through and through on the importance of positivity I finding success, which is a core element in the application of the Law of Attraction on your life. Positivity ultimately results to more positivity. Opening your mind up to the available positivity in the world will help you expand your mind to the limitless opportunities that the universe has to offer. It is all about having greater faith and also a strong belief in yourself and your capabilities.

You don't have to get stuck in a job that is not providing the much you want because you are afraid to aim higher thinking that all those other bigger jobs are out of your reach. When you have your mind filled with negative ideas you are preventing yourself from seeing better jobs that are actually just in front of you. No matter how big or small you measure your desired success, it will always be dependent on your thoughts and how you see yourself.

It is said that the moment you allow yourself to envision your deepest desires in your mind's eye and never doubt your capabilities the universe will offer you all your heart's desires no matter how big. We all have different ways in which we measure our success. Don't keep trying to succeed according to other people's definition, always work on your own idea of it and also maintain patience because success has to come to you and if you are in a hurry you will end up stressed and unhappy. It is actually much easier to attain your own definition of success.

According to the Law of Attraction success entails abundance and is not just measured in monetary value, there is more to it. The Law of Attraction helps you achieve a deeper meaning in success. It brings your way maximum happiness in relation to every aspect of your life and allows you to define life from a different more meaningful point of view. It helps us recognize the true intention of living.

The Law of Attraction gives one an opportunity for one to unlock and realize their true potential. According to this law success entails the ability to defeat all negative influences in life through the power of mind and thought. Through positive thinking we have the strength to face all trials and tribulations. Success is not only material and it comes when one realizes the true meaning of life and this realization brings with it immense happiness. This kind of happiness gives you a sense of achievement, a sense of fulfillment and wholeness.

When we understand the Law of Attraction and are receptive of the greater possibilities that the universe can offer, then we can us the power of thought to change our lives and situations into success stories. The potential and endless possibilities that lie within us are huge. If you get to know your potential then you can unlock success your way. Having a greater sense of belief in yourself and knowing what you are capable of will only be positive thinking in your personal life.

The most important thing for you to put in mind is that you are the artist of your future and it's only you with the power t make your dreams a reality. Get rid of all the negative that will deter you from painting the picture of your life that you desire. Once you have the picture in mind visualize it and start living the dream, you should be able to see yourself in the kind of state that you desire. Do not always be a person who thinks of lacking what they want rather think of them as yours and own the feeling.

LAW OF ATTRACTION FOR LOVE & RELATIONSHIP

As it has been explained, the Law of Attraction works on the basis of like attracts like. When you decide to focus on positive thoughts then you will get positive results. This law is so much applicable in love and relationships. You can use your thoughts to attract the people you love and also perfect relationships. When it comes to Law of Attraction and love, a key philosophy that is of essence is the words of Ralph Waldo Emerson who said, *"Love and you shall be loved."* What you give to the universe you will get back.

Love is our biggest preoccupations and a driving force behind our existence. Love is known to be present in all aspects of life, whether we acknowledge it or not and it is very important for our happiness and general wellbeing. It is love that breeds joy in our lives and without it life loses meaning. Many people prevent themselves from forming the kind of loving relationships they desperately long for because of ego and fear.

No matter the kind of love or relationship you are looking for, the Law of Attraction is a key solution. Many of us are standing in the way between yourself and your chances of true love which is a source of eternal happiness. It is important that you open your mind to the teachings of the Law of Attraction and the great impact it can have in your life. Our minds hold the power to open up opportunities of finding the love we crave for. As already explained, the Law of Attraction enables us to be able to bring the reality of what our hearts desire and yearn for the most, whether it is money, success, happiness or love.

Maintaining positive thoughts and discarding all negative emotions with positive affirmations of what we want in life will direct the universe to grant it all to us. The most important thing to do when you realize that there is room for more love in your life is to fill your thoughts and actions with absolute love.

It is necessary that you direct the love you feel and want to others by loving yourself and those around you. The Law of Attraction will help you find the right partner, improve your relationship with your partner and also transform your relationships with family and friends.

In order for you to attract a partner of your dreams, you should first know exactly what you want, try coming up with a specific list of the qualities you would like to attract in a mate and this will help you create awareness in your conscious and subconscious mind.

If you are in a relationship that isn't working out for you and you aren't happy about, you should first remove yourself from that situation before you start asking the universe for a new love. This will be the best way for you to start a relationship in the best way possible with no negativity. Love doesn't happen by accident, factors such as events, decisions and our manner of thinking all contribute to making love glow between two people.

It is important to understand that whether you are looking for good friendship, a life partner, good family bonds etc all these can be attracted using the Law of Attraction; all you need to do is maintain a positive mindset about everything you want.

The Law of Attraction Tips for Love and Relationships

Focus on the feelings you want

Whatever it is you want to manifest you should focus fully on it. Whether it is finding new love or improving relationships with those around you, spend some of your time thinking of what you want exactly. Think of how you will feel when you have the relationships and the kind of love you want. Embrace feelings of having those relationships and what will exist between you and those people when you have things finally in place. The Law of Attraction uses your feelings to create what you want.

Gratitude

Gratitude is an element that is so much overlooked when it comes to relationships. The longer you have stable relationships the more you start taking them for granted, forgetting how important that person or people are to you. We tend to fail expressing our feelings to those we love assuming that they already know what we feel. It doesn't matter what kind of relationship it is, always focus on the positive thoughts and be grateful for what you have.

Loving yourself

When you are able to love yourself for who you are you are telling the universe that you are worthy of love. They say we reflect to the world what we think and feel, if you can't love yourself why would you expect someone else to love you. To be loved you have to draw love. If you don't love yourself you will always have an empty feeling inside you and will be looking for something only you can provide. If you want your partner to treat you right in a relationship you should be able to show them how to treat you by loving yourself.

Enjoy life

You don't have to find love for you to start living. Life is very short and every second that passes should be highly valued. Live life to its fullest no matter what; don't wait for your soul mate to arrive first. Living happy will help you shine and stand out and will therefore attract the person who will spot you in a crowd.

Be positive

So much emphasis has been put on the fact that the Law of Attraction works only when you have the power of positive thinking. When you want love in your life then you should be positive about it and don't give up when the journey seems long. When you have failed dates and commitments then it is obvious that they were never right for you. Fix you mind on the things that you want most and trust in the power of your mind to deliver. The key to making the Law of Attraction work for you when it comes to love matters is by committing to it no matter how long it will take. Healthier and happier relationships are built by faith and strong belief in its existence.

Conclusion

Thank you again for downloading this book!

I hope this book has helped you to appreciate the Law of Attraction and embrace the ways through which it can change your life.

The next step is to practice the Law of Attraction and to get the massive possibilities that the universe has in store for you in making you achieve all that you want.

Finally, if you enjoyed this book, then I'd like to ask you for a favor, would you be kind enough to leave a review for this book on Amazon? It'd be greatly appreciated!

Click here to leave a review for this book on Amazon!

Thank you and good luck!

Printed in Great Britain
by Amazon